LOUIS PASTEUR

Titles in the
PEOPLE WHO MADE A DIFFERENCE
series include

Louis Braille
Marie Curie
Father Damien
Mahatma Gandhi
Bob Geldof
Mikhail Gorbachev
Martin Luther King, Jr.
Abraham Lincoln
Nelson Mandela
Ralph Nader
Florence Nightingale
Louis Pasteur
Albert Schweitzer
Mother Teresa
Sojourner Truth
Desmond Tutu
Lech Walesa
Raoul Wallenberg

For a free color catalog describing
Gareth Stevens' list of high-quality
children's books, call

**1-800-341-3569 (USA) or
1-800-461-9120 (Canada)**

PICTURE CREDITS
Anthony Blake Photo Library 42;
Bridgeman Art Library 9, 34-35, 43; John
Cleare 4; CNRI 53 (lower); Mary Evans
Picture Library 8, 41; Exley Photographic
Library — Nick Birch 12 (left), 16 (lower,
both), 30, 31 (upper), 35 (lower), 38, 57
(both); Giraudon 19, 26; Giraudon,
Musée des Beaux-Arts, Nantes 25;
Georges Goldner 16 (upper), 58 (lower),
59; Eric Grave 12-13, 53 (upper); Hulton
Picture Library 40 (upper), 45, 51;
Illustrated London News 11; The
Mansell Collection 55; Oxford Scientific
Films — G.I. Bernard 47 (all); Pasteur
Institute 5, 15, 36, 40 (lower), 61; Tom
Redman cover illustration; Roger-Viollet
50, 56; Ann Ronan Picture Library 21, 28,
29, 44, 46 (all), 49, 54, 58 (upper); David
Scharf 23; Science Photo Library 7, 20, 31
(lower); Sinclair Stammers 22; John
Walsh 33, 39.

North American edition first published in 1992 by
Gareth Stevens Children's Books
1555 North RiverCenter Drive, Suite 201
Milwaukee, Wisconsin 53212, USA

This edition copyright © 1992 by Gareth Stevens, Inc.;
abridged from *Louis Pasteur: The scientist who found the cause
of infectious disease and invented pasteurization*, copyright ©
1989 by Exley Publications Ltd. and written by Beverley
Birch. Additional end matter copyright © 1992 by Gareth
Stevens, Inc.

Library of Congress Cataloging-in-Publication Data

Ann Angel, 1952-
  Louis Pasteur : leading the way to a healthier world / Ann
Angel, Beverley Birch. — North American ed.
    p. cm. — (People who made a difference)
  "Abridged from Louis Pasteur: the scientist who found the
cause of infectious disease and invented pasteurization . . .
written by Beverley Birch"—T.p. verso.
  Includes index.
  Summary: A biography of the nineteenth-century French
scientist who discovered the process for destroying harmful
bacteria with heat and opened the door to the new science of
microbiology.
  ISBN 0-8368-0625-5
  1. Pasteur, Louis, 1822-1895—Juvenile literature. 2. Scientists—
France—Biography—Juvenile literature. 3. Microbiologists—
France—Biography—Juvenile literature. [1. Pasteur, Louis, 1822-
1895. 2. Microbiologists. 3. Scientists.] I. Birch, Beverley. II.
Birch, Beverley. Louis Pasteur. III. Title. IV. Series.
Q143.P2A64   1992
509.2--dc20   [B]   91-19552

Series conceived by Helen Exley
Editors: Amy Bauman, Patricia Lantier-Sampon
Editorial assistant: Diane Laska

Printed in **MEXICO**

1 2 3 4 5 6 7 8 9 96 95 94 93 92

PEOPLE
WHO MADE
A DIFFERENCE

Leading the way to a healthier world

LOUIS PASTEUR

Ann Angel

Beverley Birch

Gareth Stevens Children's Books
MILWAUKEE

The hunt for pure air

The procession wound slowly up the mountain. Guides moved ahead of a mule that swayed beneath a load of strange bottles. A small man with glasses checked the mule's harness and guided the animal along the edge of a sharp cliff.

Up and up the group climbed to the peaks of snowy Mont Blanc in the Swiss Alps. They arrived, at last, onto the unmarked snow of the Mer de Glace, which means the "Sea of Ice."

Here, a strange ceremony began. The man lit a lamp, one with a jetlike flame. He removed a bottle from the mule's back. It was a rounded flask with a straight neck tapering to a point.

The bottle contained a clear liquid that sparkled in the light. With a pair of steel pincers, the man snipped off the tip of the bottle's glass neck. Air hissed as it rushed into the narrow opening of the flask. The man quickly took the lamp and ran it across the opening. The glass melted and sealed again.

The man filled twenty flasks with the clear mountain air, sealing each again with the flame.

When he finished, he smiled with pleasure at his success.

Opposite: Pasteur went to Mont Blanc in the Swiss Alps in order to collect air samples.
Above: Pasteur collected mountain air in flasks filled with broth.

It had to be right

The day before, the man's experiments had gone wrong. He had been unable to seal the bottles. The bright sky and glaring snow had made it impossible to see the lamp's flame! Worse yet, the wind whipped so fiercely that the man couldn't aim the flame at the open neck of the flask.

Disappointed, the group had returned to the village of Chamonix. There they found a tinsmith to make a lamp that gave off a steady flame. Everything had to be done exactly right.

The great debate

On that icy morning on the Mer de Glace, Louis Pasteur was trying to prove that tiny beings called microbes lived in the air. For weeks, he and his assistants had been preparing the special bottles for their journeys. Some were carried across Paris to cellars or yards, others were placed on a hill near Pasteur's hometown of Arbois. Others were brought to the Alps for their voyage up Mont Blanc.

When Pasteur looked for proof of the microbes' existence, he left nothing to chance. "Always doubt yourself, till the facts cannot be doubted," he said. This was the secret of his enormous contribution to the world. He tested and retested, and he always paid attention to even the tiniest details.

In fact, Pasteur is said to have been involved in everything from the smallest detail to the painstaking experiments that proved the truth of the germ theory of disease. This theory says that germs — microscopic creatures known as microbes — are the primary cause of many diseases.

Although Pasteur did many experiments to prove this theory, he was not the first to recognize the existence of microbes. Scientists had known about their existence since the first microscope had been invented two hundred years earlier. But they had not thought the tiny creatures were very important.

The key to understanding disease

Pasteur proved old facts about microbes and discovered new ones. He saw how the facts were related to one another and discovered that microbes were the basis of many life processes of the world.

Pasteur saw how microbes live and die and realized that they were the key to understanding disease. His experiments helped scientists better understand the birth, life, decay, and death of matter.

Armed with this understanding, doctors were able to unlock the secrets of many diseases. With the results of Pasteur's experiments, they learned to cure and prevent illnesses that had been bothering people for thousands of years.

This fifteenth-century woodcut of a man dying from plague shows a doctor warding off evil by holding a sponge soaked in herbs to his nose. The Black Death of the 1300s killed up to half the population of Europe.

New sciences emerged in the years that followed as discovery led to discovery. The first new science was microbiology, the study of microbes. Another was the study of the cause, control, and prevention of disease. This was, and still is, done mainly through immunization and special shots called inoculations. These techniques encourage the body to develop an immunity to germs. And there was also asepsis, the practice of controlling and destroying germs in hospitals.

Pasteur and those he trained worked to revolutionize medical practice. His discoveries also rescued whole industries from disaster because of his research on silkworms, wine and beer, and cattle and sheep disease.

Right: An early picture of Paris shows the crowded and filthy conditions that allowed diseases like cholera, diphtheria, typhoid, pneumonia, tuberculosis, and the plague to spread.

Opposite: In earlier times, women died from infections, and many children died in infancy. It was common for two or three children to die in one family.

Today we take pasteurization, the technique that bears his name, for granted. Pasteurization is the process through which dairies free milk from disease-causing germs.

Controversy

Not everyone agreed with Pasteur's theories. But his work, particularly with the germ theory of disease, proved some long-held theories. He paid no attention to those who refused to accept his research or who doubted his results.

When Pasteur proved that germs live in the air during his experiment on Mont Blanc, he answered a question that scientists had been arguing about for more than one hundred years.

It is hard to think of this as a revolutionary idea. People today accept the presence of germs by using soaps and antiseptic cleaners, and by performing operations in germ-free conditions with sterilized instruments.

We now know that there are microscopic living organisms on everything in the world. Some of these microorganisms make waste materials decay to provide food for plant life. They turn raw materials into bread and wine, and some help our bodies digest food. We also know that others cause diseases when they enter the body, killing people, plants, or animals by the millions.

"I have taken my drop of water . . . full of elements most suited to the development of small beings. And I wait, I observe, I question it, I beg it to be so kind as to begin over again just to please me, the primitive act of creation; it would be so fair a sight!"
Louis Pasteur in a public lecture at the Sorbonne

None of this was known when Louis Pasteur began probing the mysteries of germs. He knew only that germs existed.

Anton van Leeuwenhoek, a Dutch scientist who died one hundred years before Pasteur was born, proved the existence of germs. Leeuwenhoek wanted to find out what things really looked like, and he had heard that a magnifying glass would help. So he designed and used the first microscope. Skin, hair, seeds, insects, and tree bark were only some of the subjects he studied. With the microscope, people quickly realized there was a world of creatures so tiny that they were invisible to the naked eye.

People burned tar and sulfur to disinfect the streets during a cholera epidemic in Granada, Spain, in 1887. On the left, a patient is being taken to the hospital. Up to five hundred people died each day during the worst of the epidemic.

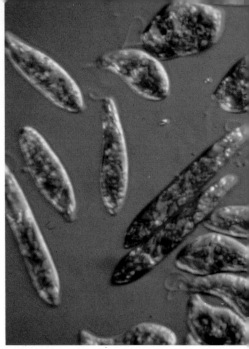

This microscope is the last one that Pasteur used to fight disease.

Peering one day through one of his magnifying lenses, Leeuwenhoek was amazed by the sight of millions of tiny creatures in rainwater from his yard! No matter how long he watched, they never stopped slithering and wriggling.

Everything he looked at was swarming with these tiny creatures. Leeuwenhoek never thought that they were causing anything, especially decay or disease. He did discover that they were killed by heat. He looked at scrapings from his own teeth and found that after drinking hot coffee, the tiny creatures were either dead or sluggish.

He recorded his findings and sent them to scientists who were members of the Royal Society in England. These

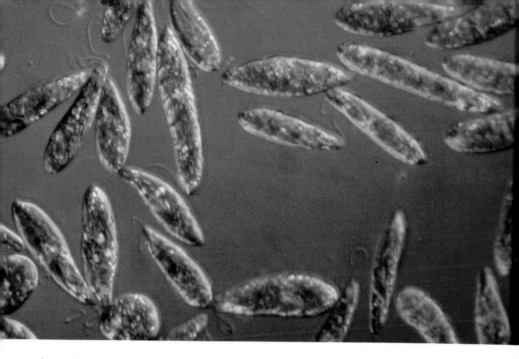

scientists repeated Leeuwenhoek's experiments and found he told the truth.

A photograph of the kind of single-celled organism Anton van Leeuwenhoek might have seen with his microscope.

Spallanzani's work on microbes

After the first excitement over Leeuwenhoek's discoveries, they were almost forgotten. In the 1850s, long after Leeuwenhoek's death, an Italian priest and professor named Lazzaro Spallanzani became fascinated by the microscopic beings.

In Spallanzani's time, there was a great debate going on. The question was this: Does every living thing have to have parents, or can living things spring into life on their own? The argument was bound up with beliefs about how the world was first formed.

Spontaneous life?

The popular idea in Spallanzani's day was that things came to life on their own. This process was called spontaneous generation. If someone buried the carcass of a bull, so the story went, out would pop a swarm of bees! Wasps, beetles, worms — animals could suddenly appear from nothing.

Spallanzani thought this whole idea was ridiculous. But how could he show he was right? One day, he read the writings of a man named Francesco Redi. Redi had proved that if flies were kept from getting to meat by covering the dish it was in, there would be no maggots!

Spallanzani experimented and showed that tiny living creatures now called microbes do not, in fact, arise spontaneously. When they appear on things or in liquids, it is because they have come there from somewhere else.

He did an experiment like the one Pasteur conducted years later on Mont Blanc to prove the same point. He boiled soups made from seeds and beans for a long time in order to kill all microbes. Then he sealed the necks of flasks filled with the soup. No new microbes could enter the flasks because of the sealed necks. This showed that microbes could not come to life in the soup or in any other substance on their own. They had to come in from elsewhere.

Microbes divide

Spallanzani read that a Swiss scientist named Horace Bénédict de Saussure had seen that microbes increased by dividing. Each one split into two, and those two split into four, and so on.

Spallanzani trapped a microbe in a drop of distilled water and watched it through a microscope until he saw it divide. The tiny rodlike shape got thinner and thinner in the middle until it became two small rods held together by no more than a cobweb-thin thread! These rods pulled apart before his eyes and became two creatures. This happened again and again.

In spite of the excitement that Spallanzani's work caused in Europe, no one learned much that was new. Interest in the creatures faded. People continued to believe in the spontaneous generation of living things.

As a student at the École Normale Supérieure, Louis Pasteur wanted to become a teacher.

The germ theory

Then came Louis Pasteur in France. His experiments dramatically changed the way scientists looked at the world. Pasteur moved scientists away from their old beliefs toward new ideas. Many methods used in medicine today grew out of Pasteur's work.

Louis Pasteur unquestionably proved that microbes travel into things from the outside. Once he had done that, he was

Above: Pasteur returned often to the town of Arbois, where he grew up in a house on the banks of the river.

Right and below: Pasteur, who loved to draw, did these portraits of his father and mother when he was a teenager. They hang in the Pasteur Institute in Paris.

able to prove the germ theory of disease. This theory stated that disease is caused by the invasion of the human, animal, or plant body by microbes that weaken it.

The scientists of Pasteur's day realized that if microbes caused many diseases, then the microbes must be tracked down and caught. Scientists needed to find ways of controlling, killing, or preventing the microbes from taking hold. Immunology, the study of how to make the body develop its own defenses against diseases, was born.

Pasteur the boy

Louis Pasteur was born to Jean-Joseph and Jeanne-Etiennette Pasteur on December 27, 1822, in the town of Dôle in France. He lived on a street that now bears his name, although in 1822 it was called rue des Tanneurs, which means "the street of the tanners." Louis's father was a tanner like many of his neighbors.

When Louis was two years old, the family moved to Marnoz. In 1827, they settled in the town of Arbois. His father sold leather from his workshop, and the family lived in a space above.

With his three sisters, Louis grew up in Arbois with its little Cuisance River running past the walls of his house and the nearby fields where small boys could play. As an adult, Louis returned to this hometown many times.

As a child, Louis showed no particular interest in science, nor was there an early clue that he would become a famous scientist. His talents seemed to be drawing and painting. At the age of thirteen, he showed a remarkable skill in painting pictures of his family and in drawing the river that ran past his house in Arbois.

To Paris

Louis developed an early ambition to become a teacher. He wanted to study at the École Normale Supérieure, in Paris. This school was founded by Emperor Napoleon Bonaparte to train professors for schools and colleges.

Louis decided to attend a boarding school in Paris that would help him prepare for the École Normale. But his first attempt at studying in Paris was disastrous. He was not yet sixteen, and the bonds that held him to his family were strong.

Louis was homesick. After about a month, his father made the journey to bring him home again. Louis returned to his drawing and painting, and he began producing a remarkable set of pastel portraits of his friends. Later, Louis went to college in Besançon, only about twenty-five miles (40 km) away.

At Besançon he began to prepare again for the École Normale. He was not

Jean Béraud, 1889

happy with his first results in the entrance exam, for although he was accepted, he placed fifteenth among the twenty-two applicants. He decided to leave Besançon, study for another year, and try again.

When Pasteur first went to Paris to study, the busy streets seemed unfriendly. He grew to like the city when he returned a year later. As a scientist and teacher, he spent most of his working life there.

Paris again

So Louis went back to Paris. This was very different from his first miserable experience. As before, he lived at the Barbet boarding school. Here he was a teacher as well as a student, rising early to give lessons to the younger boys.

It was now that the seeds of Louis's future took root. He was spellbound by

19

the scientific ideas presented to him through his education. Louis wrote excited letters home about lectures presented by his teacher and friend, Jean-Baptisté Dumas. In 1843, just before his twenty-first birthday, Pasteur entered the École Normale to learn how to teach chemistry and physics.

First explorations

When he came to the end of his studies at the École Normale, Pasteur looked around for something else to study. He wanted to become a teacher who could fire youngsters with enthusiasm, just as Professor Dumas had inspired him. To become this kind of teacher, he believed he needed to know about the smallest details of the world around him.

At this time, one of his teachers showed him a specimen of salt that had formed crystals. Although it seemed to be a pure salt, it was actually a mixture of three different shapes of crystals. Why three? Pasteur searched for the reason why nature arranged things like this.

He wondered about the most basic questions in science: What are substances made of, and can we figure out how substances are built?

Crystals and light

Crystals had attracted the attention of the curious for thousands of years. By

Crystals of table salt are magnified by an electron microscope to show their shape. Louis Pasteur's study of these shapes is called stereochemistry.

Pasteur's time, scientists knew what they looked like but not much more. A French scientist, Professor Jean-Baptisté Biot, had discovered that if a beam of light is shone through some crystals, the beam of light bends.

Pasteur wondered why. The strange things crystals did with beams of light were interesting. He wondered if there was a link between the shape of a crystal and what it did to light. He also wondered if a crystal's ingredients made a difference in how it reflected light.

He began making a careful study of beautiful crystal substances called tartaric acid and tartrates. Two forms of tartaric acid crystals were found in the crusty buildup inside wine barrels while grape juice fermented. A water solution made with the first type of crystal bent a beam of light, as Professor Biot had seen.

But if a water solution with the second type of crystal was used, it did not bend the beam of light! Yet both crystals were chemically identical, or made up of the same materials.

The first adventure

Pasteur knew he was on the verge of a discovery. He began to study crystals through his magnifying glass. He measured the angles between the different facets, dissolving them and forming them again. He struggled to

Jean-Baptisté Dumas (1800-1884), professor of chemistry at the Sorbonne and Pasteur's lifelong friend.

Tartaric acid crystals magnified eighty times.

find some difference that would explain what they did to light. The answer wasn't going to be easy to find. But he pressed on.

The first discovery

Suddenly, Pasteur found himself in the middle of his first great discovery. Peering for the thousandth time through his magnifying glass, he realized that the crystals were the same in all but one way. The facets of one crystal sloped only one way; in the other, they sloped either way.

With mounting excitement, he separated the two pieces of the second crystal. He dissolved the different pieces separately in water and predicted which way each would bend a beam of light.

His predictions were right! This showed that the structure of a crystal could be known by studying what it did to a beam of light. Investigating the behavior of a crystal showed how the crystal was built!

Pasteur's discovery suggested new ways to figure out how substances are built. His findings led to a new science called stereochemistry.

The discovery also created new questions. The crystals were alike, except that they were mirror images of each other. There was a reason for this difference, which scientists now call dissymmetry. Did nature use this difference in some way?

Pasteur was absorbed by his study for the next ten years, until an unexpected turn of events propelled him from the world of crystals.

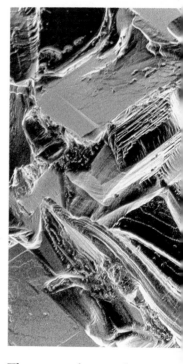

These sea salt crystals are magnified by an electron microscope. Researchers were using this microscope in the 1940s, when they first saw viruses.

Madame Pasteur

By January 1849, Pasteur had his first job as a lecturer in chemistry at the University of Strasbourg. It was here, at the age of twenty-six, that he met and fell in love with Marie Laurent, daughter of the university rector.

Shortly after their meeting, Pasteur wrote to the rector asking for Marie's hand in marriage. "My family is

"I woke up every morning with the thought that you wouldn't return my love, and then I wept! My work means nothing to me. . . ."
Louis Pasteur, in a letter to Marie Laurent before their marriage

23

comfortable but not rich," he wrote. "All I possess is good health, a willing spirit, and my work.

"I have been a Doctor of Science for eighteen months and I have presented a few works to the Academy of Sciences that have been well received," he wrote. "As to the future, all I can say is that, unless my tastes change entirely, I shall devote myself to chemical research."

He pleaded with Marie, "All that I ask, Mademoiselle, is that you will not be hasty in your judgement of me. . . . Time will show you that, under a cold and shy outside, which doubtless displeases you, there is a heart full of affection for you."

They were married on May 29, 1849. From the beginning, Marie seems to have accepted Louis's dedication to his work. She supported him, freed him from household cares, and allowed him freedom to do his research.

But she was more than a homemaker. Émile Roux, one of Pasteur's coworkers who later became famous for his own work, said Marie Pasteur encouraged her husband's work and was one of his best scientific partners.

For five years the Pasteurs lived in Strasbourg. Louis was busy with his crystals and teaching, and Marie cared for their first three children: daughter Jeanne, followed a year later by a son, Jean-Baptisté, and two years later, Cécile.

Professor of chemistry

September of 1854 brought new challenges. Pasteur was made professor of chemistry and dean of the new Faculty of Sciences in Lille, an industrial city in France known for manufacturing alcohol from beetroot juice.

Not quite thirty-two, Pasteur was young for such a responsible position. But he took his job of teaching very seriously. He wanted to fill his students with the same sense of awe in nature's miracles that he felt. Louis Pasteur's lectures were events not to be missed. He took his pupils on tours of the factories and foundries of France and Belgium.

The problem of spoiled alcohol

In 1856, a Monsieur Bigo asked for Pasteur's advice. Monsieur Bigo was an alcohol manufacturer with a business problem. Most of the time the process of changing beet sugar into alcohol in his factory was going well. But in some vats, the juice wasn't turning into alcohol; it was just turning sour. The spoiled vats were costing Monsieur Bigo thousands of francs a day.

Pasteur had some thoughts on fermentation because his studies revolved around crystals found in wine barrels during fermentation. But no one really knew anything much about fermentation except that it happened,

"Pasteur was not a trained naturalist and he was working alone, without a tradition behind him. . . ."
René Dubos, in his book
Louis Pasteur: Free Lance
of Science

Many people worked in France's wine industry, like the workers in the vineyard pictured below. Louis's invention of the process of pasteurization ensured a good product.

25

and they had known that for thousands of years.

Monsieur Bigo hoped that the man of science might have a new suggestion for his problem. So Louis went to the factory to see. He sniffed at the many vats of fermenting beet sugar. Some were filled with juice that was turning into alcohol. But other vats were filled with a slimy, sour mess. Louis couldn't find a reason for the problem. He decided he'd better have a closer look in his laboratory.

Under the microscope

Pasteur put a drop of liquid under his microscope and saw that it was filled with small, rounded shapes, or globules. Yellowish oval shapes swarmed with darker specks, smaller than any crystal he had seen. He realized these must be the yeast cells that were always present when beet sugar or grapes fermented.

Scientists did not know why the yeast cells were present. Some believed they were rotting and splitting the sugar molecules in beet or grape juice. They turned the juice into alcohol and a clear gas called carbon dioxide.

The yeast is alive!

The more he watched, the more Pasteur became convinced that the globules were alive and that they caused fermentation. He peered through his microscope and

Opposite: Louis Pasteur at work in his laboratory.

saw little buds growing on the yeasts. He understood what was happening. The yeast globules were growing and multiplying. They were feeding on the beet sugar, turning it into alcohol and carbon dioxide.

But this didn't solve Monsieur Bigo's problem with the spoiled vats. Pasteur studied a drop of the slimy stuff. There were no globules of yeast here. He picked up the bottle and took a close look. Little specks stuck to the inside and floated in the liquid.

The black rods

Pasteur checked for specks in the healthy liquid, but he found none. He placed one speck into a drop of pure water and

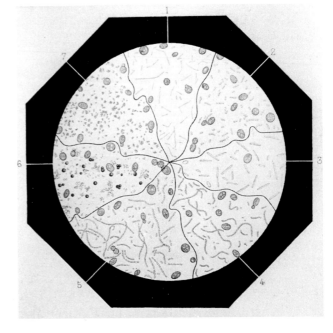

This sketch by Louis Pasteur is from his Studies on Beer, *published in 1876. It shows microorganisms that spoil wine, vinegar, beer, and milk.*

28

looked at it under a microscope. Millions and millions of tiny black rods swarmed in this single drop of water. They were busy in some kind of shimmering dance that never stopped. They were much smaller than yeast. In fact, each one could not be more than 1/25,000 of an inch (0.001 mm) long!

Pasteur studied more samples. The more the fermentation soured, the more of these creatures were in it. He finally understood what was happening. These creatures had overrun the yeast cells and prevented them from making alcohol. These little rodlike things were manufacturing lactic acid — the same substance that makes milk sour.

To be sure, Pasteur examined more samples. He discovered that whenever the vats had turned sour, the rods were there. And when they were there, there was no alcohol, only the lactic acid.

The alcohol industry is saved

Pasteur couldn't say how the rods got into the vats. But he had a strong suspicion they had come from the air.

He told Monsieur Bigo that to make good alcohol, he must test the liquid from the vats under a microscope. If he could see only the yeast cells, all would be well. But if even one rodlike creature made an appearance, he must throw the entire vat away. Once into the juice, the

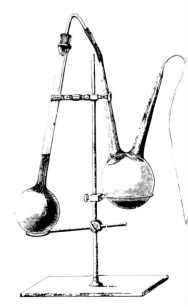

Pasteur used this contraption in his beer experiments. The left flask contained a yeast culture. Air could enter through the curved glass tube on the right without allowing microbes to travel up the tube.

black rods would multiply into millions and wipe out the yeast cells.

The alcohol industry of Lille was saved from ruin, and Pasteur was a hero. This experience in Bigo's sugar beet factory had set Pasteur on a track from which he would never be shaken. He couldn't put these creatures out of his mind. He was certain that the rods caused the lactic acid and that yeast caused the alcohol.

Microbes

Pasteur couldn't study the rods properly because they were tangled up in the pulp of the sugar beets. He had to find something the rods could grow in that would allow him to see them clearly.

He tried sugar mixed with water first. He knew that there was always sugar in fermenting liquid. In fact, he tried several mixtures but was disappointed each time when no rods grew.

Finally, he tried a broth made with yeast and a little sugar. He boiled the broth so that it would be free of any microbes. He strained the broth until it was perfectly clear.

Would the rods taken from the sick fermentation in Bigo's factory grow in this clear, germ-free yeast broth? He put a speck from a sick fermentation into a flask of the broth and carried the flask to his incubating oven. Here the mixture could be kept warm as he observed it.

The equipment Pasteur used to pasteurize beer. Scientists of Pasteur's day made their own laboratory equipment.

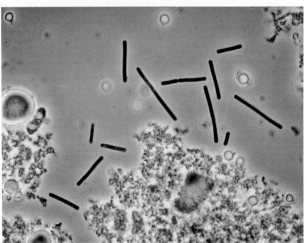

*Above: Modern wine is
pasteurized in stainless
steel tanks with
temperature controlled
by computers.*

*Left: A rod-shaped
bacteria called a bacillus
is used to produce
yogurt from milk.*

31

Throughout the day, Pasteur kept his eye on his incubator. Nothing happened. He almost became discouraged, but he was certain his guess was right. He continued to observe the oven.

And then, at the end of a long second day, he saw little bubbles of gas curling up through the broth. He squinted in the half-light, hardly daring to hope.

The bubbles were coming from the speck he had sown. And there were new specks that hadn't been there yesterday.

Millions of black rods

Pasteur put a drop of liquid under the microscope and stared with joy. Millions of rods swarmed in it! The specks had multiplied. And the same acid that had appeared in the wine vats now appeared in his yeast broth.

But he needed absolute proof.

So he took some yeast broth containing rods and put in more yeast broth, freshly boiled and free of germs. He waited again. The same thing happened. Each rod grew longer and longer. Then it split into two rods.

He put some of the rods in fresh milk. The milk soured and the rods multiplied. He did it again and again until he was certain beyond any doubt. If he added a tiny drop of rods to a clear broth, millions of new rods appeared. They always made the acid of sour milk.

The mystery is solved

Louis Pasteur had solved a mystery that had lasted for ten thousand years. Just as yeast was the cause of the fermentation that changed sugar into alcohol, these rods were the cause of the fermentation that produced lactic acid.

By August of 1857, Pasteur was absolutely certain. He read a paper about his findings to the Lille Scientific Society and told his students about fermentation. He wrote to his old teacher, Jean-Baptisté Dumas, and prepared a statement for the Academy of Sciences in Paris.

Pasteur's work on fermentation went on for many years. He showed that microbes cause fermentation in many substances. He also developed ways of preventing wine, vinegar, beer, and milk from spoiling by killing microbes with heat through the process now known as pasteurization. Today, every glass of pasteurized milk or yogurt that is free of germs is a testimony to Louis Pasteur.

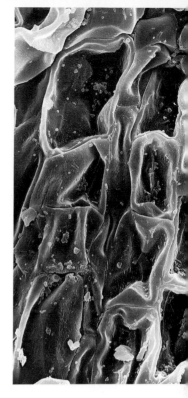

A microscope view of a wine bottle cork shows wine yeast as tiny balls against the cork cells.

Foreseeing the germ theory

In 1857, a new stage in Pasteur's life began. The École Normale in Paris asked him to become its new director of scientific studies.

Scientists no longer believed that humans, animals, or insects were produced except by parents of their own kind. But some of the scientists still

"Nothing is more agreeable to a man who has made science his career than to increase the number of discoveries. . . ."
Louis Pasteur

believed that microbes multiplied through spontaneous generation.

Everything in his fermentation work led Pasteur to believe that microbes were already in the air. They could be seen only with the help of a microscope when they landed on solids or liquids. And he was not alone in believing this. Other scientists, like his old teachers Dumas and Biot, also rejected the idea of spontaneous generation.

It was also obvious to Pasteur that fermentation and decay never took place

Above: Pasteur took walks in the city of Paris to relax from his long hours in the laboratory.

Left: The gatehouse of the École Normale, where Pasteur had his laboratory after he was appointed director of scientific studies.

Professor Jean-Baptisté Biot, Pasteur's teacher and faithful supporter.

unless microbes were present. But it was generally believed that microbes were caused by rotting. No one realized the microbes themselves caused the rotting. Pasteur's research pointed to that conclusion. But he had to prove it.

Pasteur was never happy without proof. He later said, "If I had to live my life over again, I would try always to remember [this]: 'The greatest disorder of the mind is to believe that things are so because we wish them to be so.'" He was certain that proof and only proof would move scientists in the direction he believed they must go.

Looking for proof

Pasteur knew he must find a way to show that microbes get into things from the outside. To begin with, he set up an experiment that was very much like Spallanzani's had been so many years before. He planned to kill all the microbes in a sealed container and then show that no new ones appeared. He filled several glass flasks with sugary yeast broth and boiled them to kill all microbes. As the flasks boiled, he sealed each one by melting the glass at the tip of the neck.

Then Pasteur divided the sealed flasks into two groups. He snapped off the tip of the necks of one group and allowed air in. Then he sealed them again by melting the glass. He kept all the flasks in the second group sealed. He put both groups in an oven to keep them warm enough for microbes to grow.

The results were unmistakable. In the flasks he had opened and then resealed, yeasts and other fungi were growing. In the flasks he had left sealed, nothing grew. He repeated this experiment again and again to test his idea as thoroughly as possible.

Pasteur proved that germs came only from the outside. But those who believed in spontaneous generation doubted the results. They said the microbes needed natural unheated air to burst into spontaneous life.

"There is here no question of religion, philosophy, atheism, materialism, or spiritualism. I might even add they do not matter to me as a scientist. It is a question of fact; when I took it up, I was as ready to be convinced by experiments that spontaneous generation exists as I am now persuaded that those who believe it are blindfolded."
Louis Pasteur

The final proof

Pasteur devised an experiment to prove himself right. He was convinced that it was not the air itself, but dust in the air that carried the microbes. But he couldn't find a way to let air into his flasks without also letting microbes in.

One day, Professor Antoine-Jérôme Balard, an elderly chemistry professor, strolled into Pasteur's laboratory and found him wrestling with this problem. Professor Balard agreed with Louis's germ theory and thought it would be a fine idea to prove it. He told Pasteur to prepare the flasks, then heat and bend the necks in a downward S-shape. Air

The original swan-necked flasks used by Pasteur in his experiments can be seen in the Pasteur Institute in Paris. They are still free of germs after more than one hundred years.

would pass through the necks, but dust would fall away with the force of gravity. It would not be able to travel around the bends of the glass flasks.

Louis boiled his yeast broth again, excited with this new possibility. As the liquid heated to the boiling point, the air was forced out. But now, as the liquid cooled, the air was drawn into the flasks while dust and microbes stuck in the curving neck. The fluid remained clear, without microbes. When he shook some flasks, the clear yeast broth flooded into the swan-necks, picking up the dust. These flasks became filled with microbes, multiplying rapidly!

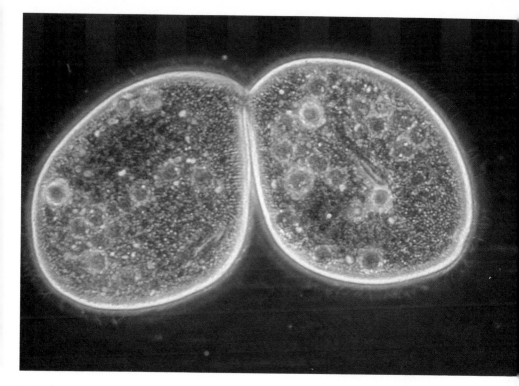

The search for pure air

Louis reasoned that the amount of dust in the air would vary in different places. Surely there would be more dust in a busy street in Paris than there would be at the top of a mountain.

Pasteur then decided to show that there were different amounts of microbes in the air in different places. He moved his experiments out of the laboratory and into the world. He and his assistants prepared sealed flasks of yeast broth. They carried ten to the cellars of the Paris Observatory, eleven more to the observatory yard, twenty up a hill near Arbois, and twenty more up Mont Blanc.

This tiny organism is in the final stages of division. Pasteur observed that many microbes increase by dividing.

"The whole secret of Pasteur's success may be summed up in a few words. It consisted in the application of the exact methods of physical and chemical research. . . ."
Sir Henry Roscoe, in the book
The Life Work
of a Chemist

Only one of ten flasks opened in the cellars developed microbes. This was because there was little dust and the air was still. In the yard, eleven out of eleven went bad. On the hill near Arbois, eight out of twenty grew microbes. And out of the twenty flasks taken up Mont Blanc, only one developed microbes.

In November 1860, Pasteur reported the results of his experiment to the Academy of Sciences: "They enable us . . . to state definitely that the dusts suspended in the atmosphere are the exclusive origin . . . [of] the existence of life in the liquids." He added, "What would be most desirable of all would be to carry these studies far enough to [do] serious research into the origin of different diseases."

Years of dedication

Over the next ten years, Pasteur disproved point after point of his opponents' objections. Some were scientific objections, but several scientists simply wished to oppose this young man who had such complete confidence.

Pasteur responded to his opponents in the best way a scientist can. He performed his experiment in public for the Academy of Sciences and proved he was right.

The 1850s and 1860s were exciting years for science. But these were also years of bitter personal tragedy for Pasteur. In September 1859, his eldest daughter, Jeanne, became ill and died suddenly from typhoid fever. She was only nine years old.

Germs everywhere

Pasteur thought the world must learn about germs. In April of 1864, he spoke at the Sorbonne to a large audience of students, authors, scientists, and ministers of state. He darkened the hall and then directed a beam of light across the room. Pasteur wanted his audience to see the millions of dust particles in the

"I cannot keep my thoughts from my poor little girl, so good, so happy in her little life who, in this fatal year now ending, was taken away from us."
Louis Pasteur, writing to his father about his daughter Jeanne's death in 1859

This 1884 engraving from the French magazine La Nature *shows Louis Pasteur working in his laboratory at the École Normale Supérieure.*

Wine being drawn from a cask in a modern winery is heated to kill microbes. This is called pasteurization.

Opposite: In the 1900s, infected bandages, used sponges, and dirt could be found in operating rooms. Because of this, many patients died from infections acquired while in the hospital.

air. He spoke to his listeners about the millions of germs drifting there.

He also showed his two flasks. One was filled with its yeast broth cloudy with microbes and the other with its yeast broth still clear. Four years had passed since the experiment.

But Pasteur did more than convert the world to the germ theory. He began acting as a wine doctor, too. The winemakers of his hometown of Arbois requested his help. Some of their wine turned sour, like vinegar, and would not keep for any length of time.

Pasteur used his microscope to study the wine. He found that there were microbes in the vats. Some were making

Scottish surgeon Joseph Lister introduced antiseptic methods into his surgery.

the wine sour, and others were making it bitter. After much study, Pasteur told the winemakers how to heat the wine just after it had finished fermenting. This would kill these microbes without damaging the wine itself. Pasteur had invented the process the world would call pasteurization.

The germ theory is applied

A Scottish doctor and professor of surgery named Joseph Lister read about Pasteur's proof of germs in the air. He was searching for a way to control infection in hospitals. He experimented with different ways to kill germs, and his work brought about many changes.

Hospitals at this time were grim places filled with the smell of blood and infected wounds. Many people died from infections they developed while in the hospital. Patients often did well the first few days after an operation, but many suddenly developed infections in their wounds and died.

In 1867, Lister began making changes. Instruments used to dress wounds were dipped in a strong solution of carbolic acid to destroy germs. Medical workers scrubbed their hands with carbolic acid. The same acid was sprayed on wounds during operations. Wounds were later washed with the solution, and Lister used antiseptic materials for dressings.

Before this, at least fifty out of every hundred patients died after surgery. Lister reduced the death rate to fifteen and then to three out of every hundred. He was thrilled with his success in reducing the death rate and grateful for Pasteur's work with germs.

The germ theory of disease, again

The question of what caused disease in the body remained open. Most doctors believed that disease was somehow in the body.

But the idea that disease and rotting were connected was an old one. Two hundred years before Pasteur, the

Pasteur's study of disease extended even to the silk industry .

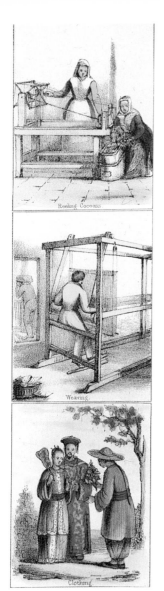

Collecting silk thread, weaving it, and wearing silk garments in China.

English scientist Robert Boyle said, "He that thoroughly understands the nature of ferments and fermentations shall probably be much better able than he that ignores them, to give us a fair account of diverse phenomena of several diseases."

Pasteur, who understood ferments, now took his first steps along the path that led to our understanding of disease.

The silkworm doctor

In 1865, Professor Jean-Baptisté Dumas asked Pasteur to go to Alais, a village in France. Pasteur was to investigate the epidemic that was killing silkworms and destroying the silk industry.

Pasteur took one of his best pupils, Emile Duclaux, and three other students from the École Normale. The Pasteur family followed later. In Alais, the pupils and the Pasteurs settled down to solve the mystery of the silkworms.

The disease seemed to start on the surface of the worms. It looked like a dusting of pepper grains. Pasteur studied hundreds of silkworms, chrysalises, and moths under his microscope. Within days, a little globule seen in the diseased insects was a sure sign of whatever was destroying them.

Pasteur told the silkworm breeders to look for globules in the moth's body after she had laid her eggs. If there were any, then the eggs were diseased and must be

Modern silkworms are raised in antiseptic conditions to prevent the diseases that Pasteur discovered in France.

Above: A silkworm moth, which lays up to four hundred eggs at a time, is surrounded by chrysalises. Silkworms hatch in ten days.

Upper left: Silkworms feed on mulberry leaves.

Lower left: A cocoon is cut away to show the chrysalis inside.

destroyed. If the moth's body was clear, the eggs would probably be sound. Healthy worms would emerge.

Disaster!

Pasteur waited until the eggs hatched in spring to find out if his predictions were right. He learned a bitter lesson. The eggs that came from healthy moths produced diseased worms! Hundreds of silkworm breeders faced disaster.

Pasteur wanted to know what had gone wrong with his theory. He returned to his microscope. Finally, the scientists found that there were two diseases at work on the silkworms. One disease was present in the globules. The other disease was a quite different microbe.

Pasteur also found out that healthy worms became sick when the droppings from sick worms soiled the mulberry leaves they ate. The second disease that had confused him, called *flacherie*, was passed on through the intestines of the worms. He had shown the importance of environment in spreading disease.

Sorrow takes a toll

Opposite: Pasteur and others were not able to find the cause of cholera. But Robert Koch found that the cholera microbe was spread through contaminated drinking water.

Disease was very much on people's minds in those years. Cholera broke out in Paris and Marseilles. As many as two hundred people a day were dying.

Disease was on Pasteur's mind in a very personal way, too. He had already

Le Petit Journal

ADMINISTRATION
61, RUE LAFAYETTE, 61

Les manuscrits ne sont pas rendus

On s'abonne sans frais
dans tous les bureaux de poste

5 CENT. SUPPLÉMENT ILLUSTRÉ **5** CENT.

23me Année ———— ** ———— Numéro 1.150

DIMANCHE 1er DÉCEMBRE 1912

ABONNEMENTS

	SIX MOIS	UN AN
SEINE et SEINE-ET-OISE	2 fr.	3 fr. 50
DÉPARTEMENTS	2 fr.	4 fr. »
ÉTRANGER	2 60	5 fr. »

LE CHOLÉRA

49

Louis Pasteur depended on his associates after a stroke in 1868 left him partially paralyzed.

lost his eldest daughter Jeanne to typhoid fever in 1859. In September of 1865, the baby of the family, his two-year-old daughter, Camille, became ill and died. Only a few months later Cécile, twelve, also caught typhoid fever. In May of 1866, she too was dead.

All this took a toll on Pasteur. Losing his children was hard enough, but he had other concerns as well. He was also troubled by his work on the silkworms. At the same time, he worried about his responsibility to people who depended on his work for their survival. On October 19, 1868, he woke up feeling a tingling sensation down his left side. By afternoon, Pasteur was shivering. During the night, his condition worsened. Before long, he could no longer speak or move. Pasteur was nearly forty-six, and he had had a stroke. His friends and family thought he was going to die.

Recovery

Louis Pasteur survived. Within weeks, he was dictating notes to his assistants. Although he was paralyzed in the left leg and arm, he refused to let this problem stop him. Within three months, he was on his way to Alais to see how the silkworm research was progressing.

He could no longer handle the scientific equipment needed for the work by himself, so he relied on his assistants to perform the experiments he devised.

The more he studied silkworms, the more certain Pasteur became about a link between the fermentations of yeasts and disease in animals and humans. Still, many doctors disagreed with Pasteur. It was Robert Koch, a doctor in Germany, who finally put an end to the doubts about Louis Pasteur's theories.

One microbe, one disease

Robert Koch lived in East Prussia in the heart of farming country. At one time, Koch had wanted to be an explorer, but he became a doctor instead.

His wife once gave him a microscope for his birthday. One day, Koch looked at some gluey black blood with his microscope. The blood had come from animals that had died of anthrax, a disease that was wiping out whole herds of sheep and cattle in Europe. Koch immediately noticed familiar rodlike microbes swarming in the blood.

He began a series of tests and experiments that proved the rodlike things were alive. They multiplied quickly inside unhealthy animals. One microbe, *Bacillus anthracis,* caused the disease of anthrax.

Koch saw that one microbe caused one disease. Pasteur had repeatedly said this, especially in the silkworm case. Koch

Robert Koch proved the principle that each specific disease was caused by one specific microbe.

"How I wish I had enough health and sufficient knowledge to throw myself body and soul into the experimental study of one of our infectious diseases."
Louis Pasteur, in a letter dated December 1873

had finally proved it. Now the hunt was on for the microbes that caused diseases all over the world.

The dawn of immunology

Pasteur was even more convinced of his theories. He was certain that there was a contest constantly being waged between microbes and the tissues they invaded.

Earlier, he had seen a cow with anthrax that had recovered from the disease. He had later seen this same cow injected with powerful anthrax bacilli. It did not die! The idea took root that having the disease somehow caused the body to develop a resistance against it.

Enough doctors recognized the impact of Pasteur's research to elect him to the Academy of Medicine in 1873. He worked harder and harder to find a way to control disease-causing microbes.

Opposite top: The anthrax bacteria, Bacillus anthracis. Anthrax can cause the death of animals and humans within hours of infection.

Opposite below: Bacillus anthracis rods in animal tissue. In humans, the disease attacks the lungs or the skin. People can catch the disease from infected animals.

The greatest breakthrough

In 1878, Pasteur studied the microbe that caused a poultry disease called chicken cholera. He grew the microbes in chicken broth. When injected into chickens, the broth culture killed them within days.

It was summertime. Pasteur and his assistants went on vacation, and a culture of chicken cholera microbes was put aside and forgotten. On returning, Pasteur injected it into a hen. The hen became sick but recovered quickly.

52

Pasteur injected more hens with the old culture. They also remained healthy. Then he injected them with a fresh culture. They were unaffected. He then injected the fresh culture into hens that had not been inoculated with the old culture. They all died.

Pasteur understood what this meant. An English doctor, Edward Jenner, had used the microbes of the disease cowpox to vaccinate people against smallpox. Jenner's method was to use a disease that was harmless to people to protect them from a disease that could harm them.

But Pasteur wanted to use microbes of the disease itself. This would increase the body's ability to fight. He called the treatment vaccination after Jenner's method. The term is still used today for this technique of preventing disease.

Pasteur wondered how many other microbes could be grown in the laboratory and used as vaccines. He spent the rest of his life looking for ways to weaken microbes' ability to multiply.

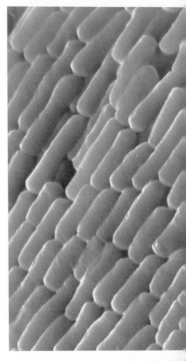

The anthrax vaccine

Pasteur also studied anthrax. Scientists knew what caused it, but ranches in France still lost thousands of cattle and sheep each year. People also died from it; an infected scratch was enough to kill.

The search for an anthrax vaccine took many years. Pasteur and his team started

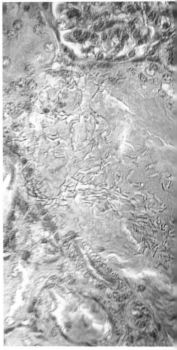

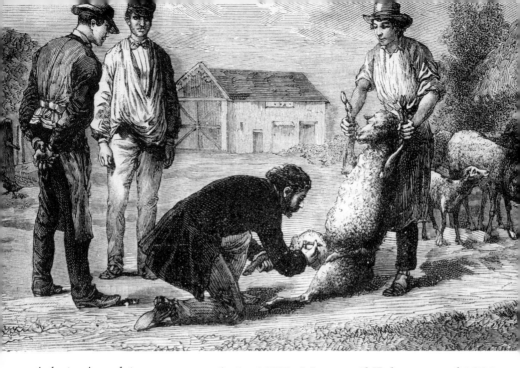

A doctor inoculates sheep against anthrax, using Pasteur's vaccine.

work in 1877. Not until February of 1881 did Pasteur believe they had succeeded with the anthrax vaccine.

The great experiment

In 1881, Pasteur accepted an invitation from farmers near Paris to test the anthrax vaccine. Animal doctors, farmers, ministers of state, scientists, and reporters gathered in May on the farm of Pouilly-le-Fort to see Pasteur and his assistants work with the sheep.

Twenty-five sheep received anthrax vaccinations. Another twenty-five sheep received no vaccinations. Both groups were kept in separate fields, and all fifty were later injected with fatal doses of powerful anthrax microbes.

"The general principles have been found and one cannot refuse to believe that the future is rich with the greatest hopes."
Louis Pasteur, speaking of the anthrax vaccine

The experiment was a huge success. All the vaccinated sheep stayed healthy, while all the unvaccinated sheep died.

Pasteur's laboratories quickly turned to making vaccines. His assistants rushed across France, injecting animals. A giant leap in medicine had been made.

Final crusade

Many people remember Pasteur more for curing rabies than for curing anthrax. Rabies was, and still is today, a disease with no cure. Persons bitten by a rabid animal, such as a dog or wolf, would begin to shake and either die from suffocation or be paralyzed.

Because people infected with rabies went into convulsions, Pasteur and his team thought that the microbe might be in the brain and spinal cord. They took fragments of infected spinal marrow from a dog that had died from rabies. They performed tests and experiments to find a rabies-causing microbe and searched for ways of weakening infected tissues to make a vaccine.

In March of 1885 Louis wrote, "I have demonstrated this year that one can vaccinate dogs or render them immune to rabies. I have not dared to treat humans bitten by rabid dogs."

The boy with rabies

On July 6, 1885, nine-year-old Joseph

Pasteur was unable to find a microbe responsible for rabies. Today, scientists know it is caused by a virus.

"Your father is as preoccupied as ever; he hardly speaks to me, sleeps little, and rises at dawn. . . . "
Madame Pasteur, writing to her children on her wedding anniversary, 1884

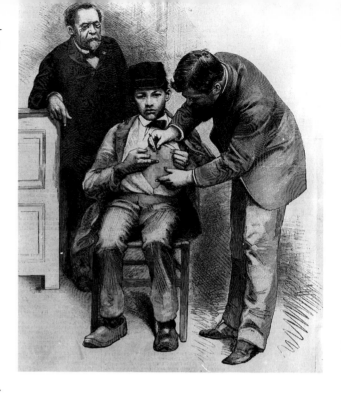

A young shepherd, Jean-Baptisté Jupille, was the second human treated with the new rabies vaccine.

Meister arrived at Pasteur's laboratory. He had been attacked by a rabid dog that bit his hands, legs, and thighs. The rabies vaccine hadn't been tested on humans yet, so Pasteur asked the advice of colleagues from the Academy of Medicine. On the evening of July 6, Pasteur supervised a doctor injecting the extract. Over the next ten days, they gave more injections, each stronger than the last. The bites healed. Joseph never contracted rabies.

News of his cure flashed around the world. People from all over Europe who had been bitten by rabid animals traveled to Paris for treatment.

Above: Pasteur's rooms at the Pasteur Institute in Paris where the scientist spent his last years.

Left: This statue of Jean-Baptisté Jupille stands on the grounds of the Pasteur Institute in Paris. The fifteen-year-old shepherd was immunized against rabies after being bitten by a rabid dog that he fought and killed in order to protect several children.

Dr. Émile Roux injected horses during his study of diphtheria.

Élie Metchnikoff worked to find ways that the body develops immunity and overcomes infection.

Eventually, the Academy of Sciences founded an institute, called the Pasteur Institute, for the treatment of rabies.

The final years

Pasteur continued working until he was nearly seventy. In 1887, when he was sixty-four years old, another stroke stopped him from doing experimental work. But he continued to lecture. In 1888, the Pasteur Institute officially opened, and he was able to witness continued research in the enthusiastic spirit that had marked his life.

On September 28, 1895, at age seventy-two, he died surrounded by family, colleagues, and students. But his spirit lived on in the scientists and doctors who inherited the knowledge he left behind.

Pasteur's legacy

The young men Pasteur trained went on to new discoveries and new glories. Among these men were Dr. Émile Roux and Dr. Alexandre Yersin. Yersin developed a treatment for diphtheria, and he also discovered the microbe that causes the plague. Élie Metchnikoff began to uncover ways that the body develops a natural resistance to microbes, called immunity.

Louis Pasteur once said to his fellow scientists, "You bring me the deepest joy that can be felt by a man whose

invincible belief it is that science and peace will triumph over ignorance and war . . . that the future will belong to those who will have done most for suffering humanity."

One wonders, thinking of Pasteur's life, if his last words to Madame Pasteur as she offered him water — the words, "I can't" — perhaps signaled the first time he entertained the possibility of failure.

Joseph Lister greets Louis Pasteur (center) during the celebration of Pasteur's seventieth birthday. Pasteur is leaning on the arm of the French president.

"You have raised the veil that for all the centuries made infectious illnesses a dark mystery."
Joseph Lister to Louis Pasteur at his jubilee

The work continues

Events moved swiftly following Pasteur's pioneering work against disease in the 1870s and 1880s. By the end of the century, most of the bacteria that caused common diseases had been identified, many of them in Pasteur's laboratory in France and in the laboratories of Robert Koch in Germany.

Even before Pasteur's death in 1895, the first evidence of a disease caused by a virus had accumulated, although it was not until the 1940s that viruses were first seen with an electron microscope. The diphtheria toxin had been identified by Émile Roux and Alexandre Yersin in 1888 at the Pasteur Institute. The tetanus toxin had been identified by scientists in Robert Koch's laboratory in 1890. Antidotes to these toxins soon followed.

In the twentieth century, we now have vaccines against smallpox, tuberculosis, yellow fever, poliomyelitis, cholera, measles, typhoid, whooping cough, rubella, influenza, and the bubonic plague.

The major leap of the twentieth century has been the development of antibiotic drugs. After the first use of penicillin in 1941, other antibiotics were developed. This is a fitting culmination to the hundred years since Louis Pasteur's work on fermentation set scientists' feet on the march against disease.

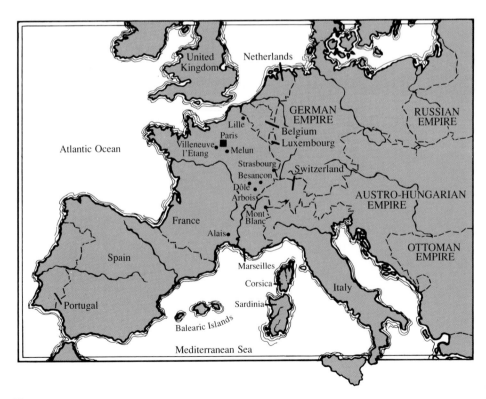

To find out more . . .

Organizations

The organizations listed below will provide you with more information about the natural sciences. When you write, be sure to tell them exactly what you would like to know. Always include your full name, address, and age. Also include a self-addressed, stamped envelope.

Young Scientists of America
  Foundation
P.O. Box 9066
Phoenix, AZ 85068

World Health Organization
Avenue Appia
CH-1211
Geneva 27, Switzerland

U.S. Center for Disease Control
1600 Clifton Road NE
Building 1, Room 2167
Atlanta, GA 30333

Books

The following books will help you learn more about Louis Pasteur and other people and issues in the health field. If you can't find them in your local library or bookstore, ask if they can be ordered for you.

Bacteria and Viruses. Leslie LeMaster (Childrens Press)
Breakthrough: The True Story of Penicillin. Francine Jacobs (Dodd)
The First Woman Doctor. Rachel Baker (Scholastic)
How Did We Find Out about Germs? Isaac Asimov (Walker and Co.)
Louis Pasteur. Rae Bains (Troll)
Louis Pasteur: Young Scientist. Francene Sabin (Troll)
The Smallest Life Around Us. Lucia Anderson (Crown)
The Value of Believing in Yourself: The Story of Louis Pasteur. Spencer
  Johnson (Oak Tree)
Viruses. Alan Nourse (Franklin Watts)
Your Immune System. Alan Nourse (Franklin Watts)

List of new words

anthrax
A disease found in animals such as cattle and sheep that can be passed on to humans. It is almost always fatal, unless vaccine is immediately available.

bacillus (plural: bacilli)
Any one of a large group of one-celled organisms that multiply by dividing in two. Bacilli are so tiny that they can be seen only with a microscope. Some cause disease or decay.

carbolic acid
A poisonous, colorless acid used as an antiseptic and a disinfectant.

cholera
A serious illness of the intestines caused by drinking water containing the cholera microbe. Humans with the disease have severe diarrhea and stomach cramps and can die from it.

chrysalis
The pupa of a butterfly or moth living in a case or cocoon.

cocoon
A covering that a butterfly or moth weaves around itself for protection during the chrysalis stage. The silkworm moth's cocoon is the source of silk fiber that is used to make clothing.

crystal
A tiny piece of a transparent mineral, such as quartz. Its surface has smooth faces or facets, arranged in a pattern.

culture
As a verb, this word means to grow microorganisms in a nutritious liquid or gelatinlike substance. As a noun, it refers to the growth that results.

diphtheria
A serious disease caused by a bacillus that produces a poison in a person's bloodstream. Extreme cases of diphtheria may result in heart failure, paralysis, or even death.

dissymmetry
The kind of likeness that exists between two objects when one is exactly the same as the other, except that it faces the opposite direction. A good example of dissymmetry is a person's hands.

fermentation
A process in which bacteria or yeasts break down the composition of plant or animal materials.

germ
A disease-causing microbe.

immunize
To make resistant to a disease, usually by shots, or inoculations.

inoculate
To place a measured dose of a weakened disease-causing microbe into a human or animal to cause the body to produce its own defense against the disease. This is also called a vaccination.

lens
A polished piece of glass that is curved on one or both sides. A lens is often used to enlarge images of small things so that their details can be observed. Such lenses are used in microscopes, telescopes, and similar instruments.

maggot
The second, or larval, stage of a fly.

microbe
Another name for microorganism. A germ is a type of microbe.

pasteurization

A process, pioneered by Louis Pasteur, of heating substances, especially liquids such as wine, beer, milk, and similar products to prevent harmful bacteria from ruining the process of fermentation.

plague

A disease caused by a bacillus and spread by fleas from infected rats. In the 1300s, the bubonic plague, or Black Death, killed up to half the population of Europe.

pneumonia

An infection of the lungs in which the lungs partly fill with fluid, making it difficult to breathe. Pneumonia is caused by a virus or by various types of bacteria.

rabies

A disease of the nervous system in warm-blooded animals that is caused by a virus. It is passed on to humans by the bite of an infected animal. Infected creatures and humans experience a heavy flow of saliva and convulsions. Humans also develop throat spasms that prevent them from drinking water. Before Pasteur developed his vaccine, death was the usual result of a rabid animal's bite.

smallpox

An easily spread and deadly disease in humans caused by a virus. Infected persons experience a cough and a rash that causes scabs to form. When the scabs fall off, they leave permanent pits in the skin. In 1796, English doctor Edward Jenner pioneered a vaccination against the disease.

spontaneous generation

A scientific theory stating that, under the right circumstances, life would appear on its own, without any cause. Louis Pasteur proved that this theory was wrong and that life always comes from other life.

stereochemistry
A branch of chemistry that studies how the atoms of a molecule are arranged. Louis Pasteur's discovery of dissymmetry in crystals gave rise to this new science.

stroke
The popular name for the blockage or breaking of a blood vessel in the brain. The results may include loss of speech, paralysis, brain damage, or even death.

tuberculosis
An infectious disease caused by a bacillus that mainly attacks the lungs. It can be acquired by drinking milk from infected cows or by breathing the bacilli that are coughed by other humans. These bacilli are killed by sunlight. But they can live in dust for weeks under damp and dark conditions. The pasteurization of milk and better sanitation are the main reasons this disease is no longer as common as it used to be.

typhoid fever
An infectious disease caused by contamination with the typhoid fever bacillus. Humans with typhoid fever experience a high fever, a rose-colored rash, and stomach pains. Two of Louis and Marie Pasteur's daughters, Jeanne and Cécile, died from this disease.

virus
An infectious microorganism. Viruses are considered parasites and can only reproduce within the body of another animal or other living being. There are many diffierent types of viruses, and they are responsible for many types of diseases. At present, there are no cures for viral diseases, so immunization remains the best protection.

yeast
A type of fungus used in fermentation and in baking. Some kinds of yeasts can cause disease in animals and humans.

Important dates

1822   **December 27** — Louis Pasteur is born in Dole, France.

1843   Pasteur enters the École Normale Supérieure.

1848   **May** — Pasteur reads his paper on crystals to the Academy of Sciences.

1849   **May 29** — Pasteur marries Marie Laurent.

1850   The Pasteurs' first daughter, Jeanne, is born.

1851   Their son, Jean-Baptisté, is born.

1853   The second Pasteur daughter, Cécile, is born.

1854   At thirty-one, Pasteur is made professor of chemistry and dean of the new Faculty of Sciences at Lille.

1856   Pasteur begins his studies of fermentation.

1858   The Pasteurs' third daughter, Marie-Louise, is born.

1859   The Pasteurs' oldest daughter, Jeanne, dies from typhoid fever. Pasteur begins studies of spontaneous generation.

1862   Pasteur is elected to the Academy of Sciences.

1863   The last Pasteur child, Camille, is born.

1864   **April** — Pasteur demonstrates the presence of germs in the air at the Sorbonne in Paris.
        Summer — Pasteur goes to Arbois, France, to test wine fermentation.

1865   **June** — Pasteur goes to Alais, France, to investigate a disease that is killing the silkworms there.

September — Camille Pasteur dies after a long illness.

1866   **May** — Cécile Pasteur dies from typhoid fever. She is buried at Arbois, as were her two sisters earlier.

1867   **July** — Pasteur is officially recognized at the Exposition Universelle for his work on pasteurization.

1868   **October 19** — Pasteur, forty-five, has a stroke. Although paralyzed on his left side, he refuses to stop working.

1873   Pasteur is elected to the Academy of Medicine.

1879   Pasteur discovers how to immunize against disease using weakened microbes.

1881   **June 2** — Pasteur's bold experiment of vaccinating sheep against anthrax is judged a complete success.

1885   **July 6** — Pasteur decides to vaccinate Joseph Meister, the first person ever to be vaccinated against rabies.

1887   Pasteur has another stroke.

1888   **November 14** — The Pasteur Institute is officially opened.

1892   **December 27** — A great ceremony is held at the Sorbonne to recognize Pasteur's achievements.

1894   The Pasteur Institute makes a vaccination for diphtheria.

1895   **September 28** — Pasteur dies at the age of seventy-two.

Index